INTRODUCTION to COMPUTERS

A Student's Guide to Computer Learning

Ms. Shikha Gupta
Ms. Shikha Nautiyal

V&S PUBLISHERS

Published by:

V&S PUBLISHERS

F-2/16, Ansari road, Daryaganj, New Delhi-110002
☎ 23240026, 23240027 • *Fax:* 011-23240028
Email: info@vspublishers.com • *Website:* www.vspublishers.com

Regional Office : Hyderabad
5-1-707/1, Brij Bhawan (Beside Central Bank of India Lane)
Bank Street, Koti, Hyderabad - 500 095
☎ 040-24737290
E-mail: vspublishershyd@gmail.com

Branch Office : Mumbai
Jaywant Industrial Estate, 1st Floor–108, Tardeo Road
Opposite Sobo Central Mall, Mumbai – 400 034
☎ 022-23510736
E-mail: vspublishersmum@gmail.com

Follow us on:

© **Copyright:** *V&S* PUBLISHERS
ISBN 978-93-815885-3-6
Edition 2019

Printed at : Param Offsetters, Okhla, New Delhi–110020

PUBLISHER'S NOTE

Introduction to Computers is ideal for first-time computer users. Written and presented in plain and simple terms, using no computer jargon, this book is an easy-to-follow manual for the absolute beginner. The information is set out in an easy to understand, step-by-step format, with clear illustrations and detailed explanations to accompany each action.

The book is divided into various chapters which explain how to use your new computer system, how computers work, how to connect all the pieces and parts, and how to start using them, how to use your keyboard and mouse, what happens when you first switch on your computer, making files and folders, the basics of Word-Processing, writing letters, creating presentations; doing fun stuff like listening to music, watching videos, editing your photos; and doing online stuff like searching for information on Google, sending and receiving e-mails, chatting with friends, etc.

So, readers, in this computerised world of today, it's time to improve your confidence and produce professional results.

PREFACE

Computer Science deals with computers and their applications. Irrespective of geographical, economical and social parameters, computers are an integral part of our lives. The pace of technical progress makes it essential that children familiarise themselves with constant computer usage. So it is imperative that a basic knowledge of computers is imparted to children as early as possible along with enough scope for them to experiment and explore applications in this medium with an ease.

Although many textbooks or reference books are dedicated to learners of high school or computer professionals, it appears that the elementary or middle school learners have been left out in the dark without a dedicated book. Keeping this in mind, we present before you **Comprehensive Computer Learning – Introduction to Computers**.

The book is an effort made with an interactive and hands-on approach to communicate the essential aspects of computers. The book introduces middle school learners to the basic technology concepts and skills through hands-on activities. Efforts have been made to keep the page count down without sacrificing important concepts and tasks. Some of the special features of the book are:

- Assessment Exercises
- Computer Trivia
- Did You Know?
- Hands-on Activity
- Let's Dwell
- More to Learn
- What it Means

Using a simple language throughout the book, technical terms have been used with careful regard to the level of comprehension. Interesting fun characters in the book make the learning all the more a fun process.

Author

CONTENTS

Hello friends, we all are familiar with the word computer and are fascinated to know about it. In this book we will help you understand interesting features related to computers. Computers and Information Technology (IT) has a deep inter-relation and let us now explore the world of IT and computers.

WHAT IS INFORMATION TECHNOLOGY (IT)?

Information Technology is a modern term that describes the combination of traditional computer and communication technologies. Revolutions in the field of computers and communications transformed the computers synonymous to Information Technology (IT).

What it means

Information is knowledge communicated concerning some particular fact, subject or event. Communication is the transfer of thoughts and message as contrasted with transportation of goods and persons

COMPUTER AND IT

Computers and information technology involves two aspects: computer competency and computer knowledge. Computer competency refers to acquiring computer-related skills. These skills are indispensable tools for today. Computer knowledge is a deeper understanding of how technology works.

Technology has certainly changed the way we live. Undoubtedly, technology plays an important role in every sphere of life. Thanks to the manifold positive effects of technology, the fields of education and industry have undergone a major change and sure, they have changed for the better.

Did You Know?

The first phase of Verbal Communication was origination of languages and the second phase was written communication (record and store the information). Some examples are: Cave paintings, clay tablets, ink, papyrus, etc. The third phase was the printing era which began with Gutenberg and his Bible in 1456.
The fourth phase started with the age of telecommunications which began with Morse's telegraph and was perpetuated by Marconi's wireless.
Now we are in fifth phase where communication is playing a major role in **interactive communication systems.**

The Latest Developments in Technology

Technology means different things to different people, and everyone has their own set of preferences when it comes to such things. There are plenty of areas where one can look for the **newest gadgets**, and here we try and provide a detailed analysis in all of these fields.

Mobile technology

The cell phones that we use have certainly come a long way since the first wireless models that started hitting the stores in the 1980s. These phones have become smaller as time has passed by, and now suddenly they seem to be getting bigger again.

1.　　**Smart-phones** perform a variety of computing tasks, in addition to your regular telephony services, and these are mini computers that are extremely advanced today. All the tasks that you perform on your PC today can also be performed on your cell phone, and this is a fact that is not missed on people.

2.　　Using the latest **3G (soon to be 4G)** wireless network technology, we can even watch live videos on our mobile phones and download media in an instant. Today, the major players in the market offering latest developments in technology are **Symbian** from Nokia, **RIM** (the makers of Blackberry phones), **Android** from Google and the **iOS** from Apple Computers.

Fig. 1: Smart-phones

Television technology

Today we view our content on HDTV (High-definition television) sets. These offer us tremendous amounts of TV resolutions and they have taken TV viewing to whole new levels. The flat screen TVs that we can buy in the market today are also built to consume very low levels of power and this has made them very energy efficient. For someone looking for some great television sets, plasma TVs, LED TVs and LCD TVs offer some great choices to pick from. These TV sets are excellent for playing video games as well and the home theater systems.

Special Feature

Video game technology: Video games are something that appeals to people of all ages, and this is a fact that has been exploited completely by developers, who regularly dish out the latest developments in technology. The graphics that we see on our video games today have certainly come a long way. Whether you are playing PC games, or if you are playing on the Sony PlayStation 3, the Microsoft Xbox 360, the Nintendo Wii, or handheld devices like the Sony PSP (to play PSP games) or the Nintendo DS, you will find a range of games that are highly engrossing.

Gaming Consoles

Online games: All the gaming consoles have their dedicated online gaming platforms where you can play against the best players from all around the world. The **PS3** has taken it a level further, by also acting as a **Blu-Ray player**.

Fig. 2: iPhone

All the latest developments in technology are supposed to help us make our lives easier, but sometimes it feels like our lives are all the more complex because of this. The importance of technology cannot be undermined in our lives, but we need to know when to draw a line under it.

SIGNIFICANCE OF COMPUTER FOR STUDENTS

Computers and the Internet technology have revolutionised the field of education. The importance of technology in schools cannot be ignored. In fact, with the onset of computers in education, it has become easier for the teachers to render knowledge and for the students to grasp it. The computer technology is used to add a fun-element to education. And it goes without saying that the Internet has endowed education with interactivity.

Computers as an interactive audio-visual media
- Visual effects provided by the animation and presentation software.
- Overhead projectors and screens facilitate a simultaneous viewing of information.

This underlines the importance of computer teaching against textbooks.

Online education and distance learning:
Online education has given a new dimension to the field of education and higher learning.
- Students do not necessarily need to be physically present in classrooms.
- Many educational institutes offer online courses to their students.
- Most of the schools and colleges offer online assignment submission facilities.
- Students can submit their homework and test assignments through the Internet.
- Many universities offer online education programs wherein the students can interact with their teachers over the web, access reference materials from the University website and earn degrees online.

What it means

Virtual Education: Virtual Education refers to instruction in a learning environment where teachers and students are separated by time or space or both and the teacher provides course content through course management applications, multimedia resources, the Internet, videoconferencing, etc.

More to Learn

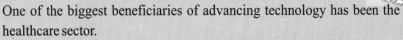

One of the biggest beneficiaries of advancing technology has been the healthcare sector.

- Communication between patients and doctors has become easier, more personal, more flexible and more sensitive.
- Personal records of patients are maintained, which makes it easier to study symptoms and carry out diagnosis of previously unexplainable conditions.
- Several medical aids have helped people overcome many medical conditions which they had to live with earlier. New medicines have led to the demise of many illnesses and diseases.

Hence friends, we can conclude that the rapid developments in Information Technology brought revolutionary changes in information processing, storage, dissemination and distribution and became a key ingredient in bringing-up great changes in over all aspects of society.

ASSESSMENT EXERCISES

1. The term refers to the widespread use of mobile communication devices:

    a. IT   b. wireless revolution

    c. PDA   d. RAM

2. A computer refers to acquiring computer related skills

    a. competency   b. aware

    c. connectivity   d. networked

3. Features of Online Education:

    a. ....   b.   ....

    c. ....   d.   ....

4. .... perform a variety of computing tasks, in addition to your regular telephony services.

5. **Examples of Gaming Consoles:**

Unit End Project:

Make a report on the basis of differences between Desktop, Laptop, Notebook and Tablet.

Social Networking: Let's dwell

The perfect gadget for **sharing photos** with the computer-less is the digital picture frame. This is a device about the size and shape of an ordinary picture frame. It contains an LCD screen, which displays multiple photos in a slide show format, and it connects to the Internet via phone line to download new pictures and information to display on the screen. Several companies have made digital picture frames, including Polaroid, Kodak, Ceiva and GiiNii. The frames are fairly similar in construction, though they do offer some different features.

Friends, now we will learn about the history and evolution of computer. Different calculating devices like Abacus, Napier's bones, Pascaline, Jacquard's Loom and Analytical Engine was evolved. In the chapter different generations of computer are described in brief.

The history of computer could be traced back to the effort of man to count large numbers. This process of counting of large numbers generated various systems of numeration like, Babylonian system of numeration, Greek system of numeration, Roman system of numeration and Indian system of numeration. Out of these the Indian system of numeration has been accepted universally. It is the basis of modern decimal system of numeration (0, 1, 2, 3, 4, 5, 6, 7, 8, 9).

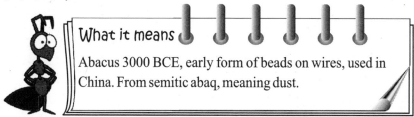

What it means

Abacus 3000 BCE, early form of beads on wires, used in China. From semitic abaq, meaning dust.

Later we will know how the computer solves all calculations based on decimal system. However, you will be surprised to know that the computer does not understand the decimal system and uses binary system of numeration for processing. We will briefly discuss some of the path-breaking inventions in the field of computing devices.

CALCULATING MACHINES

It took generations for early man to build mechanical devices for counting large numbers.

Abacus

The first calculating device called **Abacus,** was developed by the Egyptian and Chinese people.

The word 'Abacus' means calculating board. It consists of sticks in horizontal positions on which are inserted sets of pebbles. It has a number of horizontal bars each having ten beads. Horizontal bars represent units, tens, hundreds, etc.

Table Abacus

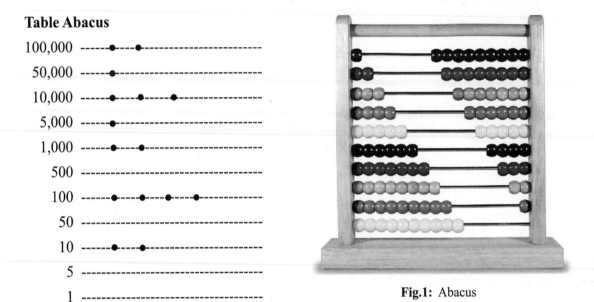

Fig.1: Abacus

Napier's Bones

English mathematician **John Napier** built a mechanical device for the purpose of multiplication in 1617 AD. The device was known as Napier's bones.

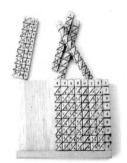

Fig.2: Napier's bones.

Computer Trivia

The original definition of Computer was any person who performed computations or was required to compute data as a regular part of their job function. Throughout history several man-made devices, such as the Abacus and Slide Rule, have been built to aid people in calculating data.

Slide Rule

English mathematician **Edmund Gunter** developed the slide rule. This machine could perform operations like addition, subtraction, multiplication and division. It was widely used in Europe in 16th century.

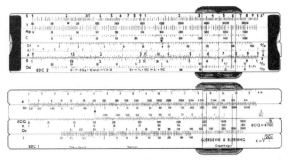

Fig. 3: Slide Rule

Pascal's Adding and Subtraction Machine

You might have heard the name of **Blaise Pascal.** He developed a machine at the age of 19 that could add and subtract. The machine consisted of wheels, gears and cylinders.

Fig. 4: Pascal's Adding and Subtraction Machine

Did You Know?

The first practical typewriting machine was conceived by three American inventors and friends – Christopher Latham Sholes, Carlos Glidden and Samual W. Soule, who spent their evenings tinkering together.

Babbage's Analytical Engine

It was in the year 1823 that a famous English man **Charles Babbage** built a mechanical machine to do complex mathematical calculations. It was called difference engine. Later he developed a general purpose calculating machine called analytical engine. You should know that Charles Babbage is called the Father of computers.

Charles Babbage

Fig. 5: Babbage's Analytical Engine

Mechanical and Electrical Calculator

In the beginning of 19th century the mechanical calculator was developed to perform all sorts of mathematical calculations and it was widely used till 1960. Later the routine part of mechanical calculator was replaced by electric motor. It was called the electrical calculator.

Fig. 6: Mechanical and Electrical Calculator

Modern Electronic Calculator

The electronic calculator used in 1960s was run with electron tubes, which was quite bulky. Later it was replaced with transistors and as a result the size of calculators became too small.

The modern electronic calculators can compute all kinds of mathematical computations and mathematical functions. It can also be used to store some data permanently. Some calculators have inbuilt programs to perform some complicated calculations.

Fig. 7: Modern Electronic Calculator

COMPUTER GENERATIONS

The evolution of computer started from **16th century** and resulted in the form that we see today. The present day computer, however, has also undergone rapid changes during the last fifty years. This period, during which the evolution of computer took place, can be divided into five distinct phases known as Generations of Computers. **Generations of computer** means the technological evolution over the period of time. Computers are classified as belonging to specific "generations". The term generations was initially introduced to distinguish between different hardware technologies.

Gradually it shifted to both hardware and software as the total system consists of both of them.

Computer Trivia

The first speech recognition software named, "Hearsay" was developed in India in 1971.

First Generation Computers (1946-59)

First generation computers used Thermion valves or Vacuum tubes. These computes were large in size and writing programs on them was difficult.

Fig. 8: First generation computers

More to Learn

Bill Gates and Paul Allen formed Traf-O-Data in 1971 to sell their computer traffic-analysis systems.

Some of the computers of this generation were:

ENIAC

It was the first electronic computer built in 1946 at University of Pennsylvania, USA by **John Eckert** and **John Mauchly.** It was named **Electronic Numerical Integrator and Calculator** (ENIAC). Today our favourite computer is many times as powerful as ENIAC, still the size is very small.

EDVAC

It stands for **Electronic Discrete Variable Automatic Computer** and was developed in 1950. The concept of storing data and instructions inside the computer was introduced here. This allowed much faster operation since the computer had rapid access to both data and instructions. The other advantage of storing instruction was that computer could take logical decisions internally.

Other Important Computers of First Generation

EDSAC

It was developed by M.V. Wilkes at Cambridge University in 1949.

UNIVAC-1

Eckert and **Mauchly** produced it in 1951 by Universal Accounting Computer setup.

Followings were the major drawbacks of first generation computers:

- They were quite bulky
- The operating speed was quite slow
- Power consumption was very high
- It required large space for installation
- They had no operating system
- The programming capability was quite low

Second Generation Computers (1959-64)

Around 1959 an electronic device called Transistor replaced the bulky vacuum tubes in the first generation computer. A single transistor contained circuit produced by several hundred vacuum tubes. Thus the size of the computer got reduced considerably. Transistors therefore provided higher operating speed than vacuum tubes. They had no filament and require no heating. Manufacturing cost was also very low. It is in the second generation that the concept of Central

Fig. 9: Second Generation Computers

Processing Unit (CPU), memory, programming language and input and output units were developed. The programming languages such as COBOL, FORTRAN were also developed during this period.

Some of the computers of the Second Generation were:

IBM 1920

Its size was small as compared to First Generation computers and mostly used for scientific purpose.

IBM 1401

Its size was small and used for business applications.

CDC 3600

Its size was large and used for scientific purpose.

The salient features of this generation computers were:

- Relatively faster than the first generation computers
- Smaller than the first generation computers
- Generated lower level of heat
- More reliable
- Higher capacity of internal storage

Third Generation Computers (1964-71)

The third generation computers were introduced in 1964. They used Integrated Circuits (ICs). These ICs are popularly known as Chips. A single IC has many transistors, resistors and capacitors built on a single thin slice of silicon. So it is quite obvious that the size of the computer got further reduced. Some of the computers developed during this period were IBM-360, ICL-1900, IBM-370, and VAX-750. Higher-level language such as BASIC (Beginners All purpose Symbolic Instruction Code) was developed during this period.

The features of computers belonging to this generation were:

- Relatively very small in size and using operating system
- High processing speed
- More reliable
- Power efficient and high speed
- Use of high level languages
- Large memory
- Low cost

Fig. 10: Third Generation Computer

Fourth Generation Computers (1971 onwards)

The present day computers that you see today are the fourth generation computers that started around 1975. It uses Large Scale Integrated Circuits (LSIC) built on a single silicon chip called microprocessors. Due to the development of microprocessor it is possible to place computer's central processing unit (CPU) on a single chip.

Fig. 11: Fourth Generation Computer

These computers were called microcomputers. Later Very Large Scale Integrated Circuits (VLSIC) replaced LSICs. These integrated circuits are so advanced that they incorporate hundreds of thousands of active components in volumes of a fraction of an inch. Thus the computer, which was occupying a very large room in earlier days, can now be placed on a table. The personal computer (PC) that you see in your school is a Fourth Generation Computer.

The salient features of this generation computers are:
- Very fast
- Very low heat generation
- Smaller in size
- Very reliable
- Negligible hardware failure
- Highly sophisticated

Fifth Generation Computers

The computers, which can think and take decisions like human beings have been characterised as Fifth generation computers and are also referred as thinking machines. The speed is extremely high in fifth generation computers.

Fig. 12: Fifth Generation Compute

Abacus and Teaching

Abacus is widely used in teaching young children. Young children are not fully aware of the number system and find it very difficult to solve arithmetical problems. They feel comfortable when they calculate using abacus. In the modern days children use abacus as their playing tool; therefore when they are taught to solve problems on their playing toy they find it fun and learn eagerly.

Some of the calculations are really difficult to perform for children. Simple arithmetic problems might feel easy to the grown up mind but they create a lot of difficulties for children and it could be done with more accuracy and precision with the help of abacus.

Abacus has also been proved beneficial for teaching blind children how to perform arithmetical functions. Its structure with the rods and beads on those rods help blind students to understand the process of calculations and allow them to calculate things themselves.

❯ TYPES OF COMPUTERS BASED ON MEMORY SIZE AND IT

On the basis of size of memory computers can be divided into following categories:

Microcomputer

Microcomputer is at the lowest end of the computer range in terms of speed and storage capacity. Its CPU is a microprocessor. The first microcomputers were built of 8-bit microprocessor chips. The most common microcomputer is a personal computer (PC). The PC supports a number of input and output devices. An improvement of 8-bit chip is 16-bit and 32-bit chips. Examples of microcomputer are **IBM PC, PC-AT**.

Fig. 13: Microcomputer

Minicomputer

This is designed to support more than one user at a time. It possesses large storage capacity and operates at a high speed than a microcomputer. The mini computer is used in multi-user system in which various users can work at the same time. This type of computer is generally used for processing large volume of data in an organisation. They are also used as servers in Local Area Networks (LAN).

Fig. 14: Minicomputer

Mainframe Computer

This type of computers is generally 32-bit computers. They operate at very high speed, have very large storage capacity and can handle the workload of many users. They are generally used in centralised databases. They are also used as controlling nodes in Wide Area Network (WAN). Example of mainframes are: **DEC, ICL** and **IBM 3000 series.**

Fig. 15: Mainframe Computer

Super Computer

This is the fastest and most expensive machine. It has high processing speed compared to other computers. They also have multiprocessing technique. One of the ways in which supercomputers are built is by interconnecting hundreds of microprocessors. Supercomputer is mainly used for weather forecasting, biomedical research, remote sensing, aircraft design and other areas of science and technology.

Examples of supercomputers are: **CRAY YMP, CRAY2, NEC SX-3, CRAY XMP** and **PARAM from India.**

Fig. 16: Super Computer

So friends, here we have learnt all about history of computer. We also learnt about its generation by generation growth and also types of computers.

ASSESSMENT EXERCISES

1. This is the first calculating device:
    - a. Pascal
    - b. Abacus
    - c. Charles Babbage
    - d. None of the above

2. This generation of computer used Vacuum tubes:
    - a. Third generation
    - b. Second generation
    - c. Fourth generation
    - d. First generation

3. It was developed by M.V. Wilkes at Cambridge University in 1949:
    - a. UNIVAC-1
    - b. EDVAC
    - c. EDSAC
    - d. ENIAC

4. Its CPU is a microprocessor:
    - a. Microcomputer
    - b. Minicomputer
    - c. Mainframe computer
    - d. All of the above

5. Select the odd one out:
    - a. ENIAC
    - b. Abacus
    - c. Slide Rule
    - d. Analytical Engine

6. ....is known as the father of computer.

7. Analytical Engine was invented by............................

8. Give definition:
    - a. Microcomputer

        ....

    - b. ENIAC

        ....

Unit end Project

1. Collect some pictures of calculating devices and paste them in your scrapbook.

2. Make your own abacus using beads and wire.

History of Social Network Sites: Let's dwell

The first recognisable social network site was launched in 1997. SixDegrees.com allowed users to create profiles, list their Friends and, beginning in 1998, surf the Friends lists. Each of these features existed in some form before SixDegrees, of course. Profiles existed on most major dating sites and many community sites. AIM and ICQ buddy lists supported lists of Friends, although those Friends were not visible to others. Classmates.com allowed people to affiliate with their high school or college and surf the network for others who were also affiliated, but users could not create profiles or list Friends until years later. Six Degrees was the first to combine these features.

Hello friends, now let us start with same basic concepts regarding computers and learn something more about its applications which include at personal level and commercial level.

Computer is a man made electronic machine. It needs electricity to work. Computer has many parts and features. It helps us in performing various activities. Computer has become an important part of our life and in fact an important part for the entire world.

What it means

The word "computer" comes from "compute" which means to calculate. So a computer is normally considered to be a calculating device that can perform arithmetic operations at enormous (great, large, huge) speed. Computer is truly an amazing machine. A few tools let you do so many different tasks.

Difference of a computer and other calculating device, like a calculator:

Calculator	Computer
Calculators are numeric tools only	Computers use numbers, words, images or sounds
Calculators are **small and cheap**er devices	Computers **are big and expen**sive devices

Fig. 1: Calculator

Fig. 2: PDA

There are various uses of computers at home level or personal level.

⟩ USES OF A COMPUTER – PERSONAL LEVEL:

A computer can be used for many simple tasks like:

Drawing Pictures: We can draw pictures on a computer.

Listening to Music: We can enjoy music on computer.

Watching Cartoons: We can watch and enjoy cartoons on a computer.

USE OF A COMPUTER AT COMMERCIAL LEVEL

Now we will learn about various applications of computer. We can see computers in various places. Few examples are given below:

IN BANK

- Many people use banks to keep their money safe.
- Banks use computers to store a list of people and their accounts.
- Banks enable online enquiry of a customer's account balance.

Computer Trivia

An ATM (Automatic or Automated Teller Machine) is a computerised machine designed to dispense cash to bank customers without the need of human interaction.

Hands on Activity

Here are some places where computers are used. Find them all!

A	Y	T	R	D	X	K	H	B
S	T	A	T	I	O	N	O	A
H	N	W	S	M	I	H	S	N
S	C	H	O	O	L	N	P	K
O	U	S	G	J	Z	X	I	C
O	F	F	I	C	E	B	T	T
A	X	S	H	O	P	V	A	A
A	I	R	P	O	R	T	L	I

In Schools and Libraries

- Computers assist teachers in teaching their students easily and quickly.
- A library uses computers to store lists of books and their location.
- Librarians use computers to issue books to you.
- Computers are used to prepare examination papers, circulars, notices, etc.

- Students use computers to prepare projects, learn various subjects and get new information.

In Shops and Supermarkets

- Shops and supermarkets use software, which calculate bills.
- They keep stock lists and list of things required.

In Medicine and Health Care

In medicine, computers are used for everything:

- Computers are used to maintain patient records and diagnosis charts for references.
- Computerised machines help in radio-diagnostics, lab diagnostics and various advanced tests to monitor and detect a patient's illnesses.

MULTIMEDIA AND TECHNOLOGY

- Computers are used to make animations, cartoons and 3-D technology movies. Animation is a technique of giving a film or a character the appearance of movement.
- Computers help in giving special effects, sound effects and colour effects to a normal graphic.

Did You know?

The first Indian 3D movie was "My Dear Kuttichathan" (Malayalam), which released 27 years ago and was later dubbed in Hindi as "Chota Chetan".

DEFENCE AND MILITARY

Computers are used for defence purposes of a company:

- Many weapons are controlled by computers
- Computers help in aircraft and air traffic control
- Various security systems use computers
- Computerised technology is used in communication and surveillance

IN SCIENCE AND TECHNOLOGY

- Computers help in research and collection of data.
- Sending space missiles and rocket launchers.
- Weather forecasting and recording and forecasting natural calamity intensity and occurrence respectively.
- Robotics – Robots are controlled by software. Robots are used in those fields which are dangerous for human beings.
- Robots help explore new planets and space bodies.

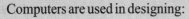

Special Feature

Computers in Information and Communication

For communication of information and messages we use telephone and postal communication systems. Computer technology helps in data and information transfer throughout the world. The Internet is a network of networks. Millions of computers all over the world are connected through the Internet. Computer users on the Internet can contact one another anywhere in the world. If your computer is connected to the Internet, you can connect to millions of computers. You can gather unlimited information through the Internet.

ASSESSMENT EXERCISES

1. A computer is used for:

    a. Listening to music   b. Playing games

    c. Internet use   d. All of the above

2. We can do calculation with the help of:

    a. Thermometer   b. IPods

    c. Video games   d. Computers

3. In railway stations computers are used for:

    a. Loading luggage   b. Ticket checking

    c. Ticket booking   d. None of the above

4. Find out the correct word:

Automatic Teller Machine	Automated Telly Machine
Electricity	Elecritcity
Calibrating device	Calculating device
Aminations	Animations

5. Give definition:

    a. ATM ....   b. Calculator ....

Unit End Project

- Make a list of few advanced technology systems which use computer technology.
- Become a detective: Visit a bank, supermarket or multiplexes. Collect evidences showing that they use computers for their functioning.

Multimedia: Let's dwell

Multimedia is nothing but the processing and presentation of information in a more structured and understandable manner using more than one media such as text, graphics, animation, audio and video. Thus multimedia products can be an academic presentation, game or corporate presentation, information kiosk, fashion-designing, etc. Multimedia systems are those computer platforms and software tools that support the interactive uses of text, graphics, animation, audio, or motion video. In other words, a computer capable of handling text, graphics, audio, animation and video is called multimedia computer.

Chapter 4

MY COMPUTER

Hello friends, in this chapter you will learn about various interesting components of a computer, its parts and the concept of I-P-O. It is also important to know about the tips related to the care of a computer and its parts.

COMPUTER – A MACHINE

A computer is an electronic machine. It is made up by joining various parts which are called hardware. Hardware is the physical part of the computer, which you can see and touch.

The four most important parts are – CPU, Monitor, Keyboard and Mouse.

Fig. 1: A Computer System

Computer Parts – Basic

Parts	Function	How does it look
CPU	CPU (Central Processing Unit) is the most important part of a computer. It does all the tasks that we want the computer to do. It controls all the other parts, such as the Monitor, Keyboard and Mouse. These parts have to be connected to the CPU, for them to work.	

Monitor	A monitor looks like a television screen. The CPU uses the monitor to show us photos, movies, games, and other activities. The mouse pointer can be seen on the monitor. Often, whatever we type using the keyboard, can be seen on the monitor.	
Keyboard	A keyboard is used to write on the computer. We can also use a keyboard to give instructions to the computer. The buttons on the keyboard are called 'keys'.	
Mouse	A mouse is used to select an item shown on the monitor. The mouse usually has two buttons and a small wheel between the buttons.	

More to Learn

Keys on keyboard

Mouse parts

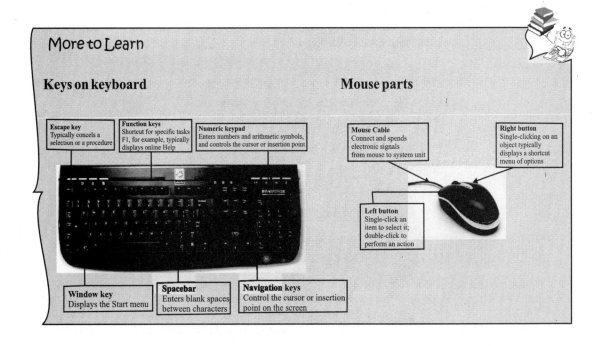

Escape key
Typically concels a selection or a procedure

Function keys
Shortcut for specific tasks F1, for example, typically displays online Help

Numeric keypad
Enters numbers and arithmetic symbols, and controls the cursor or insertion point

Mouse Cable
Connect and spends electronic signals from mouse to system unit

Right button
Single-clicking on an object typically displays a shortcut menu of options

Left button
Single-click an item to select it; double-click to perform an action

Window key
Displays the Start menu

Spacebar
Enters blank spaces between characters

Navigation keys
Control the cursor or insertion point on the screen

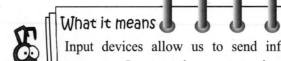

What it means

Input devices allow us to send information to the computer. Input to the computer is usually given by clicking the mouse or typing on the keyboard.

Output devices allow us to receive information from the computer. Output from the computer may be printouts from the printer, sounds on the speaker or images on the monitor.

<u>**Computer parts – Optional**</u>

Input devices

a. Scanner: The keyboard can input only text through keys provided in it. If we want to input a picture the keyboard cannot do that. A scanner is an optical device that can input any graphical matter and display it back. The common optical scanner devices are Magnetic Ink Character Recognition (MICR) **Optical Mark Reader (OMR)** and Optical Character Reader (OCR).

Scanner

- ❀ Magnetic Ink Character Recognition (MICR): This is widely used by banks to process large volumes of cheques and drafts. Cheques are put inside the MICR. As they enter the reading unit the cheques pass through the magnetic field which causes the read head to recognise the character of the cheques.

- ❀ Optical Mark Reader (OMR): This technique is used when students have appeared in objective type tests and they had to mark their answer by darkening a square or circular space by pencil. These answer sheets are directly fed to a computer for grading where OMR is used.

- ❀ Optical Character Recognition (OCR): This technique unites the direct reading of any printed character. Suppose you have a set of hand written characters on a piece of paper. You put it inside the scanner of the computer. This pattern is compared with a site of patterns stored inside the computer. Whichever pattern is matched is called a character read. Patterns that cannot be identified are rejected. OCRs are expensive though better than the MICR.

b. Joystick: The joystick is a vertical stick which moves the graphic cursor in a direction the stick is moved. It typically has a button on top that is used to select the option pointed by the cursor. Joystick is used as an input device primarily used with video games, training simulators and controlling robots.

Joystick

Did You Know?

QWERTY Keyboard was discovered by Christopher Latham Sholes in 1870, his aim was to slow down typing speed and prevent jams in types.

Output Devices

a. Printer: It is an important output device which can be used to get a printed copy of the processed text or result on paper. There are different types of printers that are designed for different types of applications. Depending on their speed and approach of printing, printers are classified as **impact** and **non-impact printers.**

❀ **Impact printers** use the familiar typewriter approach of hammering a typeface against the paper and inked ribbon. Dot-matrix printers are of this type.

❀ **Non-impact printers** do not hit or impact a ribbon to print. They use electrostatic chemicals and ink-jet technologies. Laser printers and Ink-jet printers are of this type. This type of printers can produce colour printing and elaborate graphics.

Printers

b. Plotter: Plotters are used to print graphical output on paper. It interprets computer commands and makes line drawings on paper using multi-coloured automated pens. It is capable of producing graphs, drawings, charts, maps, etc. Computer Aided Engineering (CAE) applications like CAD (Computer Aided Design) and CAM (Computer Aided Manufacturing) are typical usage areas for plotters.

Plotter

c. Audio Output: Sound Cards and Speakers: The Audio output is the ability of the computer to output sound. Two components are needed: Sound card – plays contents of digitised recordings; Speakers – attached to sound card.

I-P-O CYCLE

I-P-O means input-process-output cycle. The I-P-O cycle is a general cycle which we come across in our daily life, even in living world.

A plant takes in water and gives flowers. Taking in is called Input. Giving out is called Output. Water is the input and flowers are the output for the plant.

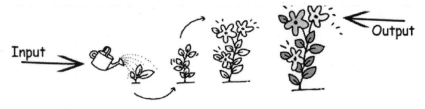

A computer has four major functions:

Input	Accepts data
Processing	Processes data
Output	Produces output
Storage	Stores results

What it means

Data: Data is words, numbers, dates, images, sounds, etc., without context. Data is the raw facts given to the computer. Data items need to be part of a structure, such as a sentence, in order to give them meaning.

Information: Information is a collection of words, numbers, dates, images, sounds, etc., put into context, i.e. to give them meaning.

A computer is only useful when it is able to communicate with the external environment. When you work with the computer you feed your data and instructions through some devices to the computer. These devices are called Input devices. Similarly a computer after processing gives output through other devices called Output devices.

In simple terms, input devices bring information INTO computer and output devices bring information OUT of a computer system. These input/output devices are also known as peripherals since they surround the CPU and memory of a computer system.

* Keyboard and mouse are input devices
* Monitor, speakers and printer are output devices
* Any device has to be connected to the CPU in order to receive or send information. CPU controls all the connected input and output devices. The CPU is neither an input nor an output device. It takes the information given by the input devices, does the work and sends the result to the output devices. This is called **Processing.**

▷ PROCESSING AND STORAGE DEVICE

Till now we have learnt about input and output devices. The knowledge is incomplete if we do not know about the features of processing device and storage devices.

Process unit: The task of performing operations like arithmetic and logical is called **processing**. The takes data and instructions from the storage unit and makes all sorts of calculations based on the instructions given and the type of data provided. It is then sent back to the storage unit.

Computer Trivia

In computer's memory both programs and data are stored in the binary **form.**
The binary system has only two values 0 and 1. These are called bits. **As human** beings we all understand decimal system but the computer can only understand binary system. A number of switches in different states will give you a message like this: 110101....10. So the computer takes input in the form of 0 and 1 and gives output in the form 0 and 1 only.

Functional Unit of a Computer

Arithmetic Logical Unit (ALU)

After entering data through the input device it is stored in the primary storage unit. The actual processing of the data and instruction are performed by the Arithmetic Logical Unit (ALU). The major operations performed by the ALU are addition, subtraction, multiplication, division, logic and comparison. Data is transferred to the ALU from storage unit when required. After processing the output is returned to the storage unit for further processing or getting stored.

Control Unit (CU)

The next component of computer is the Control Unit, which acts like the supervisor seeing that things are done in proper fashion. The control unit determines the sequence in which computer programs and instructions are executed. It also acts as a switch board operator when several users access the computer simultaneously. Thereby it coordinates the activities of the computer's peripheral equipment as they perform the input and output. Therefore it is the manager of all operations performed by the ALU.

Central Processing Unit (CPU)

The ALU and the CU of a computer system are jointly known as the central processing unit. You may call the CPU as the brain of any computer system. It is just like the brain that takes all major decisions, makes all sorts of calculations and directs different parts of the computer functions by activating and controlling the operations.

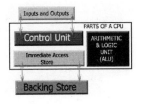

Memory/Storage system of a computer

Computer memory is used to store two things: i) instructions to execute a program and ii) data. When the computer is doing any job, the data that have to be processed are stored in the primary memory. This data may come from an input device like keyboard or from a secondary storage device like a floppy disk/CD/DVD.

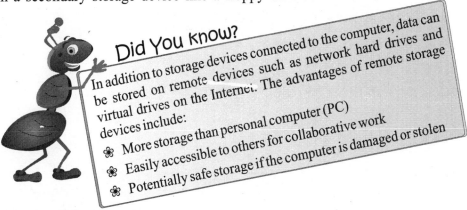

Did You know?

In addition to storage devices connected to the computer, data can be stored on remote devices such as network hard drives and virtual drives on the Internet. The advantages of remote storage devices include:

❀ More storage than personal computer (PC)

❀ Easily accessible to others for collaborative work

❀ Potentially safe storage if the computer is damaged or stolen

Main/Primary Memory

The most common type of memory that most users are familiar with is 'main/primary memory' or **'RAM'** (random-access memory).

Random access memory (RAM) is an area in the computer system unit that temporarily holds a user's data, operating system instructions and program instructions.

The other forms of primary memory are: ROM (Read Only Memory), PROM (Programmable Read Only Memory), EPROM (Erasable Programmable Read Only Memory), etc.

Magnetic disks, Floppy disks, Optical Disks and external USB drives are the **permanent way to store information**, found both internally and externally on most personal computer systems. Examples of storage devices include hard disk drives, floppy disk drives, CD-writers, DVD drives and flash drives. Each storage device (except for the hard drive) also requires a storage medium that actually holds the data. These storage medium devices include floppy disks, CDs, DVDs, flash drive and tapes.

The main storage device on a personal computer system is the **internal hard drive**. It holds both data and program files. Hard drives come in different sizes, measured in megabytes or gigabytes (billions of bytes). Sample sizes of hard drives include 40GB, 80GB, 100GB. The hard drive is generally referred to as drive C.

Hard disk

A CD (compact disk) provides 650-700 MB of storage. A DVD (digital video disk or digital versatile disk) has a capacity of 4.7 GB. CDs and DVDs are durable storage and have a higher tolerance for temperature fluctuations than hard disk, floppy disks, and tapes. They are unaffected by magnetic fields and dust and dirt can be cleaned off easily. The biggest threat is scratches.

Compact Disk

A USB flash drive is a popular portable storage device. It is about the size of a highlighter pen and is very durable. It plugs into a USB port at the back of the computer and provides fast access to data. The advantages of the USB flash drive are:

* Durability
* You can open, edit, delete, and run files just as if they were on hard drive
* It plugs into USB port
* Provides fast access to data and uses little power

Pen drive

From the following list write I for input, P for process, S for storage and O for output related term or device.

- Mouse: ....;
- Plotter: ....;
- Hard Disk: ....;
- Keyboard: ....;
- ALU: ....;
- Printer: ....;

So friends, in this unit we have learnt about various computer parts – input, process and output devices and the well known I-P-O cycle.

ASSESSMENT EXERCISES

1. A computer is another name for a notebook computer.

        a. PDA   b. Laptop

        c. Handheld   d. DVD

2. .... is sometimes referred to as temporary storage.

        a. PDA   b. CD

        c. RAM   d. DVD

3. Match the following:

A	B
CU	Display/output
MB	Clicking
ROM	Unit of data storage
Mouse	Process Unit
Monitor	Temporary Storage

4. Name three optional devices which give us output:

        ....;

        ....;

        ....;

5. Expand the following short forms:

        ALU

        EPROM

        RAM

        VDU

        PDA

Unit end Project:

Find out information related to latest tablets, PDAs and touch screens.

List them below.

....

....

....

Aakash Tablet: Let's dwell

India's **Cheapest Tablet** PC:
Our country, India launched the world's cheapest tablet computer in October, 2011, which is meant to be sold to students at the subsidised price of $35 and later in shops for about $60. The **Aakash** is aimed at university students for digital learning via a government platform that distributes electronic books and courses. The Aakash has been developed by UK-based company DataWind and Indian Institute of Technology (Rajasthan).

Friends, in this session we will' learn about starting instructions of a computer. We will also learn how to start and close it.

▷ STARTING OF COMPUTER

Electricity is a prime need of a computer. We can say that electricity is the food of a computer. Switchboards in our houses are the source of electricity power.

We have to be very careful while switching ON and OFF a computer. Let's learn:

Starting a computer system is a step-wise procedure.

Instruction to Start a Computer

- Switch on the main power supply button on the switchboard.
- Switch on the power button on the UPS.
- Switch on the power button on the CPU. An indicator glows which shows that the computer system is on.
- Press the switch of the monitor to 'ON' position.
- Taskbar will appear at the bottom of the computer screen.
- Icons will appear on the screen. A computer runs on electricity of about 240 volts.

Fig. 1: Switching ON a Computer

Monitor screen

This screen is called **desktop window**. We can see small pictures on the desktop. These are called **icons.** These icons are used to represent various functions and programs.

We can see a small icon 🌀 in the left corner. It is **the start** icon. To do **any** work on the computer you need to start a program.

Fig. 2: Monitor screen

Hands on Activity

Switch on the computer with the help of your elder and make a list of icons you see on the desktop window.

....

....

....

....

....

❯ OPERATING KEYBOARD AND MOUSE

Keyboard

It is one of the most important parts of a computer which is used to enter commands, text, numerical data and other types of data by pressing the keys on the keyboard. A keyboard is an Input device.

It is a text base input device that allows the user to input alphabets, numbers and other characters. It consists of a set of keys mounted on a board.

Alphanumeric Keypad

It consists of keys of English alphabets, 0 to 9 numbers, and special characters like + − / * ( ) etc.

Function keys

There are twelve function keys labeled F1, F2, F3…. F12. The functions assigned to these keys differ from one software package to another. These keys are also user programmable keys.

Fig. 3 Keyboard

Functions of some of the important keys are defined below. They are also called Special Function keys.

Enter key

It is similar to the 'return' key of the typewriter and is used to execute a command or program.

Spacebar key

It is used to enter a space at the current cursor location.

Backspace key

This key is used to move the cursor one position to the left and also delete the character in that position.

Delete key

It is used to delete the character at the cursor position.

Insert key

Insert key is used to toggle between insert and overwrite mode during data entry.

Tips for using your keyboard safely

- Place the keyboard at elbow level
- Avoid resting your palms or wrists on any type of surface while typing
- While typing, use a light touch and keep your wrists straight
- When you are not typing, relax your arms and hands

Shift key

This key is used to type **capital letters when** pressed along with an alphabet key. It also used to type the special **characters located on** the upper-side of a key that has two characters defined on the **same** key.

Caps Lock key

Cap Lock is **used to** toggle between the capital lock features. When 'on', it locks the alphanumeric keypad for capital letters input only.

Tab key

Tab is used to move the cursor to the next **tab position** defined in the document. Also, it is used to insert indentation into a document.

Control key

Control key is used in conjunction with other keys to provide additional functionality on the keyboard.

Alt key

Also like the control key, Alt key is always used in combination with other keys to perform specific tasks.

Esc key

This key is usually used to negate a command. Also used to cancel or abort executing programs.

Numeric Keypad

Numeric keypad is located on the right side of the keyboard and consists of keys having numbers (0 to 9) and mathematical operators (+ − * /) defined on them. This keypad is provided to support quick entry for numeric data.

Cursor Movement Keys

These are arrow keys and are used to move the cursor in the direction indicated by the arrow (up, down, left, right).

Special Feature

Keyboard Shortcuts

Press this	To do this
HOME	Move the cursor to the beginning of a line or move to the top of a webpage
END	Move the cursor to the end of a line or move to the bottom of a webpage
CTRL+HOME	Move to the top of a document
CTRL+END	Move to the bottom of a document
PAGE UP	Move the cursor or page up one screen
PAGE DOWN	Move the cursor or page down one screen
DELETE	Delete the character after the cursor, or the selected text; in Windows, delete the selected item and move it to the Recycle Bin
INSERT	Turn Insert mode off or on. When Insert mode is on, text that you type is inserted at the cursor. When Insert mode is off, text that you type replaces existing characters.

LEFT ARROW, RIGHT ARROW, UP ARROW, or DOWN ARROW: Move the cursor or selection one space or line in the direction of the arrow, or scroll a webpage in the direction of the arrow

Mouse

Mouse is an input device that is used with our personal computer. It is used to point to a particular place on the screen and select in order to perform one or more actions.

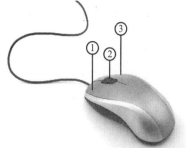

1. Primary button   2. Scroll wheel   3. Secondary button

Is a pointing device which is used to control the movement of a mouse pointer on the screen to make selections from the screen? A mouse has one to five buttons. The bottom of the mouse is flat and contains a mechanism that detects movement of the mouse.

It can be used to select menu commands, size windows, start programs, etc. When we roll the mouse across a flat surface the screen censors the mouse in the direction of mouse movement. The cursor moves very fast with mouse. Mouse gives us more freedom to work in any direction. It is easier and faster to move through a mouse.

Mouse actions

A mouse typically has two buttons: a primary button (usually the left button) and a secondary button (usually the right button). We use the primary button most often. It also includes a scroll wheel between the buttons to help us scroll through documents and webpages more easily. The scroll wheel can be pressed to act as a third button.

Left Click: Used to select an item.

Double Click: Used to start a program or open a file.

Right Click: Usually used to display a set of commands.

Drag and Drop: It allows you to select and move an item from one location to another. To achieve this place the cursor over an item on the screen, click the left mouse button and while holding the button down move the cursor to where you want to place the item, and then release it.

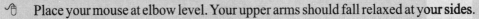

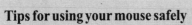

More to Learn

Tips for using your mouse safely

- Place your mouse at elbow level. Your upper arms should fall relaxed at your sides.
- Do not squeeze or grip your mouse tightly. Hold it lightly.
- Use a light touch when clicking a mouse button.
- Keep your fingers relaxed. Do not allow them to hover above the buttons.
- When you do not need to use the mouse, do not hold it.

▶ SWITCHING OFF COMPUTER

Switching off computer does not mean to switch **off the** main power supply switch. A computer must be properly switched off. It is known **as sh**ut down.

When you complete your work, click on Start icon ⬤. A list of items will appear. This list is called Start Menu.

Move the pointer to 'Shut Down' computer option and click the left mouse button.

Fig. 5 Switching Off Computer

What it means

Clicking: Clicking is most often used to select (mark) an item or open a menu. This is sometimes called single-clicking or left-clicking.

So friends, here we have finished with learning of starting of a computer, working of Keyboard and Mouse, and Switching off a Computer.

ASSESSMENT EXERCISES

1. This is an input device.

    a. Monitor   b. Keyboard

    c. Mouse   d. All of the above

2. A computer runs on electricity of about:

    a. 240 volts   b. 220 volts

    c. 140 volts   d. 120 volts

3. This key is usually used to negate a command.

    a. Esc   b. Enter

    c. Alt   d. Number

4. Select the odd one out:

    a. Monitor, keyboard, printer, mouse

    b. Keyboard, Esc, Alt, Ctrl

    c. Switching ON , Switching OFF, Shutting down, Closing

    d. Alphabet keys, Function keys, Number keys, Special keys

5. Give definition:

    a. Icon

    ....

    b. Alphanumeric Keypad

    ....

6. Give example of:

    Number keys

    ....

    Special keys

    ....

Unit End Project

1. List down various Asanas and Exercises to relax your following body parts after working on computer.

    Wrist

    Eyes

    Shoulders

    Neck

Login & Password: Let's dwell

When a computer is used by more than one person, each user is given a username and password to access it. If the username and password are not typed correctly, the dialog box prompts you to enter them again.

Hello friends, up till now we have learnt about computer, various parts (hardware) of computer and how to start a computer and various desktop features. Now in the following unit we will introduce:
- What is Software ● Notepad ● MS Paint

WHAT IS SOFTWARE?

The term software refers to the programs or instructions that enable a computer to perform its tasks. Computer software falls into two primary categories: System software and Application software.

What it means

Software: A programme which consists of step-by-step instructions that helps the computer how to do its work. Software is another name for a programme or programmes. The purpose of software is to convert data (unprocessed facts) into information (processed

System software represents programmes that allow the hardware to run properly.

Examples of System software are: MS DOS, Linux, Windows Vista, Window 7, Unix, etc.

Computer Trivia

MS-DOS stands for Microsoft Disk Operating System.

Fig. 1: System Software

Fig. 2: Application Software

Application software represents programmes that allow users to do something besides simply run the hardware.

Examples of Application Software: MS Paint, MS Office 2010, Notepad, WordPad, Google Chrome, Adobe Reader, etc.

Did You Know?

Viruses are small programmes that hide themselves on your disks (both diskettes and your hard disk).

NOTEPAD

Notepad is an application software. Notepad is a **handy programme through which a user can** type-in text quickly and easily. It is a basic text-editing program and it is most commonly used to view or edit text files.

Computer Trivia

A text file is a file type typically identified by the .txt file name extension.
.txt means when you browse for your already created document, only notepad documents (those ending with .txt) will show up in the Open window.

Opening Notepad

To open Notepad click on start button 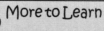 All Programs → Accessories → Notepad

Fig. 3: Notepad Window

More to Learn

Open Notepad by clicking the Start button. In the search box, type Notepad and then, in the list of results, click Notepad.

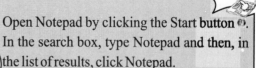

Editing features of Notepad

We can use Notepad for writing text. To enter text, just start typing in Notepad. By using the backspace key we can edit the mistakes.

Cut, copy, paste, or delete

1. To **cut text**, select the text, click the Edit menu, and then click **Cut**.
2. To **copy text**, select the text, click the Edit menu, and then click **Copy.**
3. To **paste the text** you have cut or copied, click the location in the file where you want to paste the text, click the Edit menu, and then click **Paste**.
4. To **delete text**, select it, click the Edit menu, and then click **Delete**.
5. To **undo** your last action, click the Edit menu, and then click **Undo**.

Change the font style and size

Changes to the font style and size affect all the text in the document.

1. After opening the Notepad, Click the Format menu, and then click **Font**.
2. Make your selections in the Font, Font style, and Size boxes.
3. When you are finished making font selections, click OK.

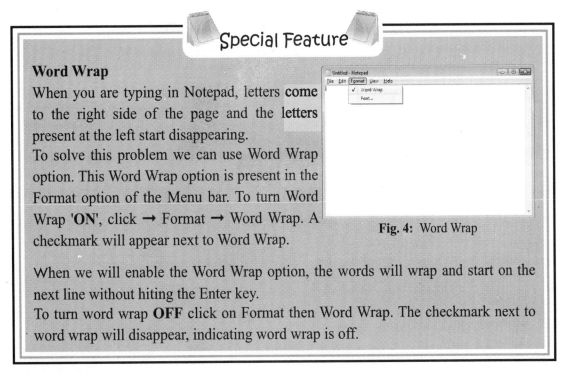

Special Feature

Word Wrap

When you are typing in Notepad, letters come to the right side of the page and the letters present at the left start disappearing.

To solve this problem we can use Word Wrap option. This Word Wrap option is present in the Format option of the Menu bar. To turn Word Wrap 'ON', click → Format → Word Wrap. A checkmark will appear next to Word Wrap.

Fig. 4: Word Wrap

When we will enable the Word Wrap option, the words will wrap and start on the next line without hiting the Enter key.

To turn word wrap **OFF** click on Format then Word Wrap. The checkmark next to word wrap will disappear, indicating word wrap is off.

Saving the Text

To save your text click on File → Save and if you have not saved it before the following screen will appear. Here you can save the file with a particular name.

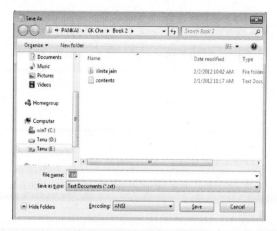

Fig. 5: Saving Notepad file

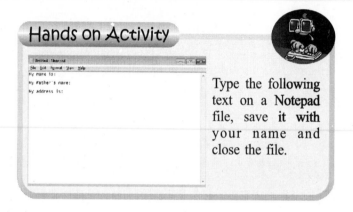

Type the following text on a **Notepad** file, save **it with** your name and close the file.

⟩ MS PAINT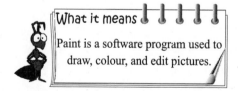

MS Paint is also an application software, it is a drawing program which can be used to create drawings or edit digital pictures.

You can use Paint like a digital sketchpad to make simple pictures and creative projects or to add text and designs to other pictures, such as those taken with your digital camera.

What it means
Paint is a software program used to draw, colour, and edit pictures.

Start MS Paint

Click on Start **button** 💠 All Programs → Accessories → MS Paint

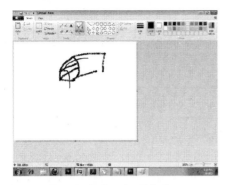

Fig. 6: MS Paint Window

MS Paint window

The Paint Window appears on the screen that provides you the Paint Area. You can increase the size of the Window by clicking on the Maximise button.

The **MS Paint Window** has four main components.

As you can see, all of the tools and buttons are neatly positioned at the upper part of the window.

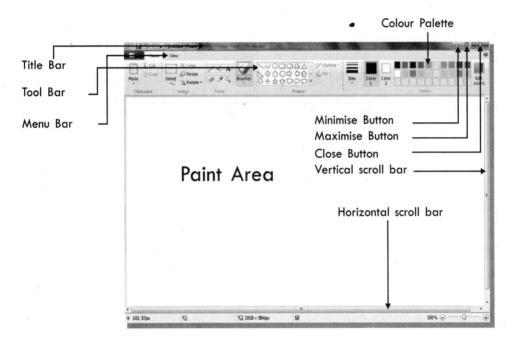

Fig. 7: Paint Window Components

Menu bar: It is used for creating, saving, opening, editing or deleting a file.

Tool box: It is located on the upper side in the ribbon. It provides drawing and painting tools used for making pictures, drawing lines and writing text.

Colour Palette: It is located at the top in the ribbon. It provides various colour options used for painting the pictures. It can be very evidently seen in the above figure.

Paint Area: It is the area where you can draw and paint.

Tool Box

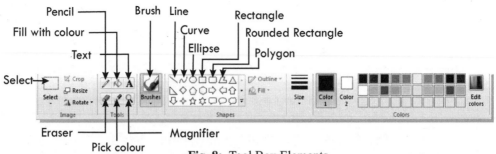

Fig. 8: Tool Box Elements

Tool	Tool icon	Functions
Free-Form select Tool		It is used to select freely any part of drawing mainly for cut/copy and paste purpose.
Selection Tool		It allows you to select a rectangular part of your drawing for cut/copy and paste purpose.
Eraser/colour Eraser Tool		It erases the portion of the drawing over which it is dragged.
Fill with colour Tool		It is used to fill the whole or a part of drawing with any colour selected from colour palette.
Pick colour Tool		This tool picks up colour from any part of the picture just by clicking over it.
Magnifier Tool		It is used to see your drawing in enlarged form.
Pencil Tool		It is used for free-hand drawing as you draw on your drawing book.
Brush Tool		It is also used for free-hand drawing but it draws with a thick outline. The shape and size of the brush can also be changed.
Airbrush Tool		Using this tool you can spray colours into your drawing.
Text Tool	**A**	This tool allows you to write text in the paint area.
Line Tool		It is used for drawing straight lines of varying thickness.
Curve Tool		It is used to draw curved lines.
Rectangle Tool		It is used to draw rectangle or square.
Polygon Tool		It is used to draw polygons.
Ellipse Tool		It is used to draw ellipse or circle.
Rounded Rectangle Tool		It is used to draw rectangle with rounded corners.

Try your hand on the Paint software and create an art work of your own. (One example is given below)

Saving Paint File

1. Take the mouse pointer to the File menu. Select "Save As" option and click the mouse button. "Save As" window appears on the screen. Enter any name in the File name box, say 'Scenery'.

2. Click on the save button. Your drawing is now saved on the hard disk under the name, 'Scenery. Bmp'.

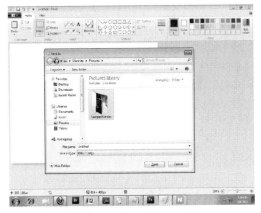

Fig. 9: Saving a Paint file

So friends, here we have learnt about Softwares and their categories – System and application softwares. We have also learnt about two simple application softwares (Notepad and MS Paint). Notepad helps us in typing text and MS Paint helps us in creating various pictures.

 ASSESSMENT EXERCISES

1. We can use text tool in:
   a. MS Word
   b. MS Paint
   c. MS Excel
   d. All of them

2. MS Office is an:
   a. System Software
   b. Application Software
   c. Virus
   d. None of them

3. Word Wrap is present in this software:
   a. MS Paint
   b. Notepad
   c. Windows
   d. None of these

4. .txt is an extension for this file type:
   a. MS Paint
   b. MS Word
   c. MS DOS
   d. Notepad

5. Select the odd one out:
   a. Linux, Unix, MAC OS, Adobe
   b. Paint, Notepad, MS Office, MS Windows
   c. Rectangle, Airbrush, Word Wrap, Line tool
   d. Google Chrome, Mozilla Firefox, Internet Explorer, MS Windows

6. Tool box is present on the .... side of the screen.

7. .... provides various colour options.

8. Give definition:
   a. Software

   ....

   b. Ribbon

   ....

Unit End Project

1. Draw a picture of a computer similar to the one used in your computer room.

2. Open a new file and draw a tree using pencil tool and colour it. Enlarge your tree by 400%.

Tux Paint: Let's dwell

Tux Paint is a free computer art software for children.
Tux Paint for Windows is available as an installer program,
downloadable as an executable program (.exe) which you can
double-click to begin the installation process.
http://tuxpaint.org/download/windows/

Friends, till now we have learnt so many things about Computer, its parts and many more. Now in this session we will learn a new thing which is called Operating System. Operating System is a bridge between human language and computer language.

OPERATING SYSTEM

An operating System is the software that makes it possible for you to work on your computer and have it perform the tasks you need. It also allows you to enter commands in the computer by clicking on objects that appear on the monitor.

An Operating System is the most important software that runs on a computer. It manages the computer's memory, processes, and all of its software and hardware. It also allows us to communicate with the computer without knowing how to speak the computer's "language". **Without an Operating System, a computer is useless.**

Types of Operating Systems

Operating Systems usually come preloaded on any computer that you buy. Most people use the Operating System that comes with their computer, but it is possible to upgrade or even change Operating Systems.

The three most common Operating Systems for personal computers are:

   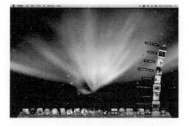   

Fig. 1: Microsoft Windows   **Fig. 2:** Apple Mac OS X   **Fig. 3:** Linux

Modern Operating Systems use a "Graphical User Interface" or GUI. A GUI lets us use our mouse to click on icons, buttons, and menus, and everything is clearly displayed on the screen using a combination of graphics and text.

Computer Trivia

Each Operating System's GUI has a different look and feel, so if we switch to a different Operating System it may seem unfamiliar at first. However, modern Operating Systems are designed to be easy to use, and most of the basic principles are the same.

Microsoft Windows, which is popularly known as MS Windows or in simple words we can say Windows is the most **popularly used Operating Syste**m.

Fig. 4: Microsoft Windows

Now we will learn more about windows and its parts, such as desktop, files and folders.

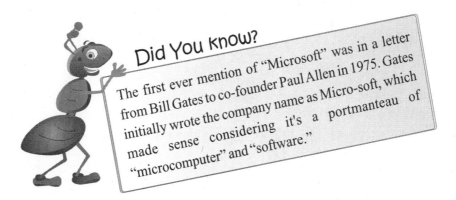

Did You Know?

The first ever mention of "Microsoft" was in a letter from Bill Gates to co-founder Paul Allen in 1975. Gates initially wrote the company name as Micro-soft, which made sense considering it's a portmanteau of "microcomputer" and "software."

⟩ DESKTOP

The Desktop is where you might keep useful things always accessible to you. When you switch on your computer and start Windows, the first screen that you see is the desktop. The Windows Desktop is somewhat like a real desktop. It's always there, though it may be recovered by other program windows.

Desktop Icons

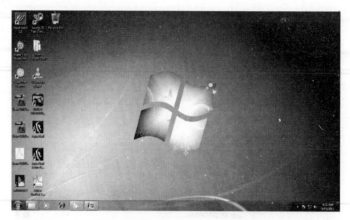

Fig. 5: Desktop picture with icons

The desktop contains small pictures with some names, such as My Computer, Documents, Control Panel, Recycle Bin, etc. These small pictures are called icons.

My Computer

My Computer is the most important Desktop item. This is where everything is present: your hard drive, floppy, CD-ROM, computer's main control (Control Panel). All the computer resources can be accessed from My Computer – it is the gateway to your computer. To access My Computer, double-click on its icon in the Desktop.

Fig. 6: My Computer icon picture

My Computer window looks just like any window you open, you can minimise or maximise this window and close by clicking on x.

My computer contains

    (i)   Hard Drive (s) (C: D:)

    (ii)   3.5 Floppy Drive (A:)

            CD_ROM Drive (E:)

    (i)   Printers

    (ii)   Control Panel

    (iii)   Dail-up Networking (used especially for Internet Connection)

    (iv)   Scheduled Tasks (give your computer jobs to perform when you are out)

In My Computer you are forbidden to Delete, Copy, Rename anything. (Try to press delete key and nothing will happen.) The items in My Computer window are so crucial that you cannot run your computer without them. For example, if you remove the hard drive where would you store your files? Since, My Computer items are related to the computer system they should remain intact.

Special Feature

Operating Systems for Mobile Devices

The operating systems that we have been talking about were designed to run on desktop or laptop computers. Mobile devices such as phones, tablet computers, and mp3 players are very different from desktop and laptop computers, so they run Operating Systems that are designed specifically for mobile devices.

Examples of mobile operating systems include Apple iOS, Windows Phone 7, and Google Android.

Fig.7: Examples of mobile Operating Systems

Besides My Computer, there are quite a few icons on the Desktop.

These are as follows:

My Documents

This is where you store all your personal stuff like files, anything you create with your computer. You can put your things somewhere else also. Think of My Documents as your Desktop drawer where you can put your papers.

Fig. 8: My Document

Recycle Bin

Provides space for deleted files from folders or Desktop, and provides a second chance to recover files deleted.

Fig. 9: Recycle Bin

Date, Time, Volume

At the bottom right corner of the screen, you can tell what time is it. By clicking your mouse over the clock icon, you'll get the date. You can also control the sound by clicking on the speaker icon.

Start icon

This icon lies at the bottom left of your screen. It provides an easy way to start different programs and other applications.

Fig. 10: Start Icon

Taskbar

This lies at the bottom of your screen. This displays all open applications and windows. The Start button and the Taskbar work together and help you in managing your applications

Taskbar

Fig. 11: Taskbar

More to Learn

Personalise Your Desktop's Background and Themes

Windows 7 has some amazing new themes and backgrounds to choose from. They include **vivid photography, digital artwork** and **Aero themes** that use colour and glass effects in an appealing way.

How to Apply

- In the Search bar of the Start Menu, type and select Personalisation.
- It will show a list of many options. It includes:

Personalisation

Change theme

Change desktop background

Change windows glass colour

Change the Screensaver

Personalise your computer

Change window colour and metrics

Set screensaver password

Adjust screen resolution

Adjust clear typed text

MANAGING FILES, FOLDER AND WINDOWS

To use Windows, you need to know two things: files and folders.

A File in Windows is a collection of data that is stored in files. Files can be of many types like: picture files, movie files, document files, sound files, etc. When you use a programme to work, you create a file. Files are basically the same data. The data they store may be different.

Each file has a name, a location and a size. Depending on the type of file, each file is denoted by an icon. Each file is given a name called 'file name' which is useful to identify the file. A file name has two parts: **a name** and **an extension.**

After working with computer for a while, you can end up with hundreds of files. So at that time you can use folders to organise your files.

Folders are also called **Directories.** It can be assumed like a boxes to put things in. The icon for a folder looks like a briefcase. A folder can also contain other folders in it.

Change your desktop background and write down the steps:

....

....

....

....

....

Managing your "Windows"

Microsoft Windows is called "Windows" for a reason. Programmes appear on your screen as "Windows" (rectangular shapes) and are laid 3-dimensionally on top of one another (see image at right), just like on a real desktop. The desktop is your work surface and all of your open windows appear on top of it.

If you can see a window, that means it is open and the programme is running. It is possible to make the window bigger, smaller, or close it by using the buttons in the top right corner of any window

Minimise:

When we Left-click this button to shrink the window down to a small button that will appear in the task bar.

Fig. 12: desktop with minimise button

Restore/Maximise:

Left-click this button to make the window as large as it can be – it should take up your entire screen.

Maximise/Restore

Fig. 13: desktop with maximise/restore button

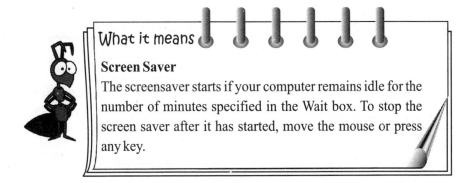

Close

When we Left-click this button, window will be closed. The program will close and stop running. Make sure you save your work first if you are typing a document.

Restore Down

When we Left-click this button, window will be smaller in size without minimising it.

Close

Fig.14: Task bar picture showing minimising, restore, close buttons

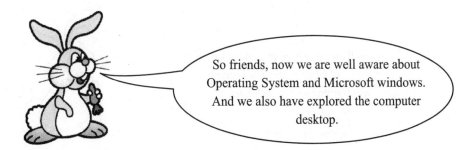

So friends, now we are well aware about Operating System and Microsoft windows. And we also have explored the computer desktop.

ASSESSMENT EXERCISES

1. This is the most important Desktop item:

    a. My Computer   b. My Documents

    c. task bar   d. title bar

2. This lies at the bottom of your screen:

    a. start icon   b. task bar

    c. recycle bin   d. My Documents

3. They work together and help you in managing your applications open or close windows easier.

    a. The title bar and the Taskbar   b. The Start button and the title bar

    c. The Start icon and the Taskbar   d. The recycle bin and the Taskbar

4. File in Windows is a collection of data that is stored in:

    a. desktop   b. taskbar

    c. files   d. computer

5. To access ...., double-click on its icon in the Desktop.

6. The .... is somewhat like a real desktop.

7. Folders are also called ....

Unit End Project

1.   Start MS windows and identify the files and folder icons and hard disk drive in my computer and make a report on your experience of this whole process.

2.   Suppose, by mistake you deleted all your important files from your desktop, now how will you recollect those files? Make a report.

Operating System: Let's dwell

Apple-OS X Lion is the world's most advanced OS, launched on Mac. OS X is engineered to take full advantage of the technologies in every new Mac. And to deliver the most intuitive and integrated computer experience. Tap, scroll, pinch, and swipe using Multi-Touch gestures, directly controlling what's on your screen in a more fluid, natural, and intuitive way. When you scroll down on your trackpad or Magic Mouse, your document scrolls down. When you scroll up, your web page scrolls up. When you swipe left, your photos move left. And there are many more gestures that make all you do on your Mac the best experience you can have on a computer.

Chapter 8

INTRODUCTION TO MS WORD

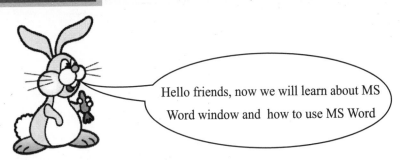

Hello friends, now we will learn about MS Word window and how to use MS Word

Computer **software refer**s to the programs or instructions that enable the computer to perform its tasks. MS Office is an integrated software package. It is developed by Microsoft Corporation (USA).

MS OFFICE

MS Office is a Graphical User Interface (GUI) software. It is powerful, efficient and user friendly. MS-Office consists of four major applications:

1. MS Word   2. MS Excel

3. MS Access   4. MS Power Point

What it means

Graphical User Interface (GUI): A GUI (usually pronounced GOO-ee) is a graphical (rather than purely textual) user interface to a computer. In GUI the user gets a more friendly approach of using and handling computer operations. Elements of a GUI includes windows, pull-down menus, buttons, scroll bars, iconic images, etc.

WORD-PROCESSING

One such computer software helps us type and write letters, words and prepare lengthy reports and helps us in doing our homework.

Word processing means to type and edit any kind of text such as letters, articles, memos, etc. Word processing involves typing, editing and printing. Here, keyboard is used as a typewriter but a keyboard has special keys that a typewriter does not have. Word Processing is one of the most widely used applications installed on a computer for the purpose of writing, editing and creating reports or documents.

The followings are examples of some popular word processor:

- Soft word
- WordStar
- Word perfect
- Microsoft Word

MS WORD

The application software MS Office has MS Word as the word processor application.

MS Word application has been evolving and Microsoft is coming out with latest versions with new applications after every few years. The latest version of MS Word available is MS Word 2010.

In this Unit we will learn many features of MS Word 2010 and basics of Word processing.

Computer Trivia

One of the main advantages of a word processor over a conventional typewriter is that a word processor enables you to make changes to a document without retyping the entire document.

Starting MS Word

➡ Click on Start icon, a start menu appears on your screen.

➡ Select and click on Programmes.

➡ Go to MS Office.

➡ Finally click on Microsoft Word.

➡ This will open a Microsoft Word Screen.

= Click on Start button All Programs → Microsoft Office → Microsoft Office Word 2007

Fig. 1: Opening MS Word

The Microsoft Word window contains the following parts:

1. Ribbon: The Ribbon displays all of the options and functions of the word applications. In order to open one of the menu options, you need to point your mouse to the item of your choice and click by using the left mouse button.

Ribbon Tabs: Click any tab on the ribbon to display its buttons and commands.

Ribbon Groups: Each ribbon tab contains groups, and each group contains a set of related commands.

a. When you open Word 2010, the ribbon's **Home tab** is displayed. This tab contains many of the most frequently used commands in Word. For example, the first thing you will see on the left side of the tab is the Clipboard group, with the commands to Paste, Cut and Copy, as well as the Format Painter.

b. Next, in the **Font group,** you will find commands to make text bold or italic.

c. It is followed by the **Paragraph group** with the commands to align text to the left, right, or centre, and to create bulleted and numbered lists

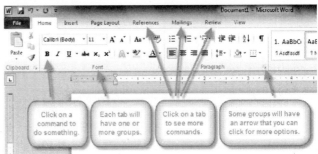

Fig. 2: Ribbon Elements

2. Backstage View: Backstage view gives you various options for saving, opening a file, printing or sharing your document.

Getting to Backstage View: Click the File tab.

➡ You can choose an option on the left side of the page.

➡ To get back to your document, just click any tab on the Ribbon.

Fig. 3: Backstage View

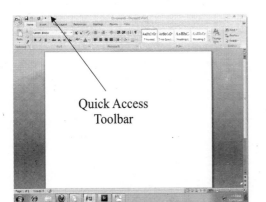

Quick Access Toolbar

Fig. 4: Quick Access Toolbar

3. Quick Access Toolbar: The Quick Access Toolbar in the upper-left corner of the Word programme window provides shortcuts to commands you will use often.

KeyTips: Word 2010 provides shortcuts for the ribbon, called KeyTips, so that you can quickly perform tasks without using your mouse. To make KeyTips appear on the ribbon, press the ALT key.

Fig. 5 KeyTips

Next, to switch to a tab on the ribbon using your keyboard, press the key of the letter displayed under that tab. In the example shown above, you would press N to open the Insert tab, P to open the Page Layout tab, S to open the References tab, and so on.

4. Title Bar: This bar is located at the top of the word screen. The Title bar displays the name of the programme and the name of the document that is currently open on your screen. **5. Ruler:** There are two rulers available, vertical and horizontal. The vertical ruler shows the length of page or document, white the horizontal ruler shows page width, position of tabs, indents, etc.

6. Status bar: It is located at the bottom of the window, which shows page number, current line and column number, etc.

7. Document Area: This is the area where you can type your text.

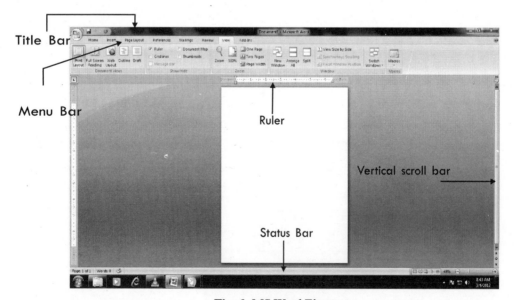

Fig. 6: MS Word Elements.

Using MS Word

Opening a new document

1. Open MS Word application as instructed.

2. Click New option of the File menu. A New dialog box will appear. Double click the Blank Document icon and press Create button. As a result, a new blank document will appear on the screen.

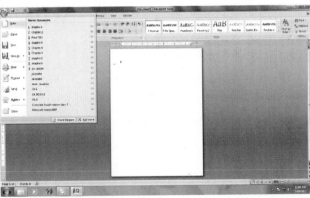

Fig.7: Opening New Document

Inserting Text

1. Move your mouse to the location where you wish text to appear in the document.
2. Click the mouse. The insertion point appears.
3. Type the text you wish to appear.

Deleting Text

1. Place the insertion point next to the text you wish to delete.
2. Press the Backspace key on your keyboard to delete text to the left of the insertion point.
3. Press the Delete key on your keyboard to delete text to the right of the insertion point.

Selecting Text

1. Place the insertion point next to the text you wish to select.
2. Click the mouse, and while holding it down, drag your mouse over the text to select it.

Release the mouse button. You have selected the text. A highlighted box will appear over the selected text.

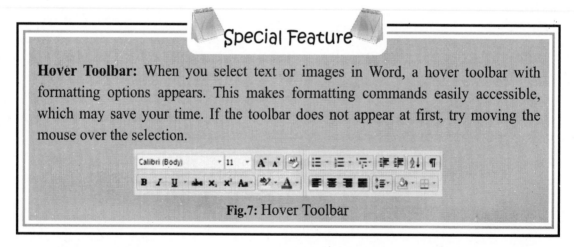

Special Feature

Hover Toolbar: When you select text or images in Word, a hover toolbar with formatting options appears. This makes formatting commands easily accessible, which may save your time. If the toolbar does not appear at first, try moving the mouse over the selection.

Fig.7: Hover Toolbar

Copy and Paste Text

Fig. 10: Cut-Copy-Paste Commands.

1. Select the text you wish to copy.
2. Click the Copy command on the Home tab. You can also right-click your document and select Copy.
3. Place your insertion point where you wish the text to appear.
4. Click the Paste command on the Home tab. The text will appear.

   Similarly you can cut particular text portion and paste it somewhere else if required.

Formatting elements

1. Font type: Choose a font which is clear and easy to read. If you mix different types of fonts in a sentence, it will be difficult to read the sentence.

2. Font size: Larger fonts are used for text which is important and to capture the attention of the reader.

3. Font Style: Whenever you want to highlight or emphasise a word or sentence:

- Use either bold or italics style of the font
- Use bold and italics together only if necessary
- A different colour can be used for a word or sentence
- Underline can also be used to draw attention
- Remember not to use too much of **bold**, italics or colour in your story, essay **or** any document that you are preparing

Fig. 11: Formatting window

Saving Text file/Document

"Save" Command

1. Click the Save command on the Quick Access Toolbar.
2. The document will be saved in its current location with the same file name.
3. If you are saving for the first time and select Save, the Save As dialog box will appear.

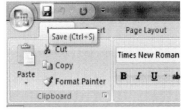

Fig. 12: Saving Document

"Save As" Command

Save As allows us to choose a name and location for your document

1. Click the File tab
2. Select Save As
3. The Save As dialog box will appear. Select the location where you wish to save the document
4. Enter a name for the document and click Save

Fig. 13: Save As document

The files in the MS Word application are saved as an extension – '.doc'

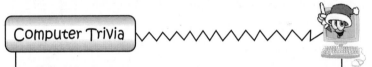

Computer Trivia

In Windows 7, files are saved into a **Documents library,** and in other versions of Windows, files are saved to the **My Documents** folder.

Closing Document

After completing the work do the following to close your document:

1. Click on Close option from File menu. It will caution you for saving your document.
2. Click on Save button. If the document is already saved the application will close, but if the work in the document is not saved, then upon selecting the save option, a Save As window will open and you have to follow the above mentioned steps.

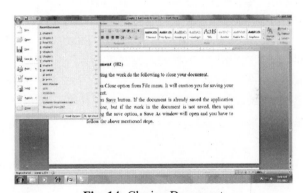

Fig. 14: Closing Document

Hands on Activity

- Open an existing Word document
- Select a sentence
- Copy and paste the sentence from one location in the document to another
- Select another sentence
- Cut and paste the sentence to another location in the document

So friends, in this unit we have learnt about the Word Processor software – MS Word and how to use MS Word.

ASSESSMENT EXERCISES

1.   The is one of the application software programmes.

    a. word processor   b. database management system

    c. Operating System   d. application software

2.   MS Office is a .... software.

3.   Give full form of:

    GUI: ....

    MS: ....

4.   A file named 'My Computer' will be saved in MS Word as:

    a. .txt; b. .bmp; c. .doc; d. .xls

5.   Give path for:

    a. Opening a new word document:

    ....

    ....

    b. Closing a word document:

    ....

    ....

Unit End project:

Prepare a document with the following specifications as under:

Heading: My parents (Font – **Arial Black**, Font Size-14)

Description: 10 lines (Font – Times New Roman, Font Size – 12, Name of your parents should be 'Bold')

The Paragraph should be justified (Ctrl+J)

Fonts should be colourful.

Components of Paragraph Tab:
Let's dwell

The Paragraph group with the commands to align text to the left, right, or centre, and to create bulleted and numbered lists.

Paragraph Tab

Chapter 9

INTERNET AND MULTIMEDIA

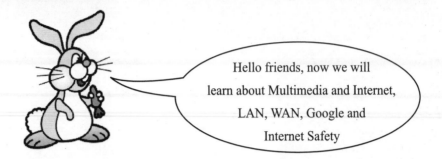

Hello friends, now we will learn about Multimedia and Internet, LAN, WAN, Google and Internet Safety

WHAT IS MULTIMEDIA?

Multimedia is a combination of video and audio with lot of graphics. Speakers, CD-ROM, digital cameras, DVDs and even video cameras provide a complete multimedia experience on a Computer System. The computer systems at home and schools when added with these multimedia components make usage of the computer more informative and recreating. Multimedia programmes are interactive and digitally defined to provide children with great learning experience.

USE OF MULTIMEDIA

Children like to work on multimedia because it is innovative and interesting as it has sound, pictures and animations.

Multimedia helps us in doing various activities like:

- Watching movies
- Listening to music
- Computer games
- Educational and interactive CD-ROMS
- Creativity by making videos

Fig.1: Multimedia Components

WORLD OF NETWORKING

Today computer is available everywhere and therefore there is a need to share data and programmes among various computers. Computer networking is basically the process of connecting two or more computers or devices, using hardware and software, so that data can be transferred and shared among them. It is just like making a chain of people who are holding hands and passing on information from one person to another.

More to Learn

Network means connecting more than one computer together through hi-speed cables. In a network information can be shared from one computer to another easily. Networking is a type of communication.

LAN

Network used to interconnect computers in a single room or rooms within a building or nearby buildings is called Local Area Network (LAN). This usually spans about 0-5 kms and is generally a private network owned by an organisation. For example: Office LAN, Hospital LAN, Campus-wide LAN, etc.

Fig. 2: Networking – LAN

WAN

The term Wide Area Network (WAN) is used to describe a computer network spanning a regional, national or global area. For example, for a large company the head quarters might be at Delhi and regional branches at Mumbai, Channai, Bengaluru and Kolkata. Here regional centres are connected to head quarters through WAN. The distance between computers connected to WAN is quite large. Therefore the transmission medium used is normally telephone lines, microwaves and satellite links. Internet is an example of a WAN.

Fig. 3: Networking-WAN

WHAT IS INTERNET?

Internet is a useful multimedia application. Nowadays Internet has become very useful informative and entertaining tool for us. "The Internet is a worldwide network of computers linked together by hi-speed **cables. Anybody who has a** computer or PC can become a part of the internet world by connecting through an ordinary telephone line."

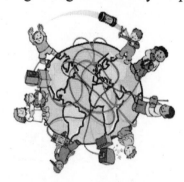

Fig. 4: Internet – Connecting world.

Origin of Internet

In the late 1960's, Department of Defence, USA launched a network of computers called the ARPANET, also called Advanced Research Project Agency Network. The main focus of this project was to share information regarding science and technology for research purposes. In 1989, with the advent of World Wide Web (www), the internet became so flexible that anyone in the world was able to connect to each other.

Uses of Internet

The Internet is useful in many ways. Throug Internet we can get information on any topic we want.

- We can send messages through internet within few seconds, which are known as E-mails.
- We can talk to people on Internet, even we can see them live on the screen of the monitor.
- We get latest updates related to global issues, instantly.

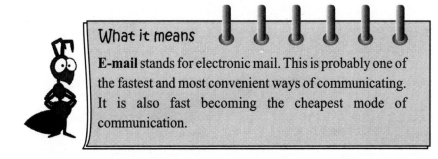

What it means

E-mail stands for electronic mail. This is probably one of the fastest and most convenient ways of communicating. It is also fast becoming the cheapest mode of communication.

▶ INTERNET SAFETY

One of the important issues is safety while using internet. The two major points while working on internet are:

Internet Security: We must secure our computers with technology in the same way that we secure the doors to our homes.

Internet Safety: We must act in ways that help protect us against the risks that come with Internet use.

More to Learn

An Internet firewall helps create a protective barrier between your computer and the Internet.

Primary Online Risks and Threats

To Computer	Viruses/ Worms	Software programs designed to invade your computer and copy, damage, or delete your data.
	Trojans	Viruses that pretend to be helpful programs while destroying your data, damaging your computer and stealing your personal information.
	Spyware	Software that tracks your online activities or displays endless ads.
To Children	Cyber-bullies	Both children and adults may use the Internet to harass or intimidate other people.
	Invasion of Privacy	If kids fill out online forms, they may share information you don't want strangers to have about them or your family.
	File-share Abuse	Unauthorised sharing of music, video, and other files may be illegal, and download viruses or worms.

Four Steps to Help Protect Your Computer

- Use an Internet firewall
- Keep your operating system up-to-date
- Install and maintain antispyware software
- Install and maintain antivirus software

Computer Trivia

What is a 'blog'?

A 'blog' (short form of web log) is an online journal where you can write about your interests, what's happening in your life, or a myriad of other things.

What is Google?

Now that you know about the Internet, the first interesting thing that one should know is Google. Google is famous as a search engine worldwide. Google roots back to 1995 when two graduate students, **Sergey Brin** and **Larry Page**, met at Stanford University. In 1996, Brin and Page collaborated on a research project that was to eventually become the Google search engine.

Google Window

Google was chosen for its resemblance to the word googol, a number consisting of a numeral one followed by a hundred zeroes, as a reference to the vast amount of information in the world. Google's self-stated mission: "To organise the world's information and make it universally accessible and useful." Google became so dominant that the name has become a verb meaning to conduct a Web search; people are as likely to say they "Googled" some information as to say they searched for it.

Hands on Activity

Open a Google page and now search the following topic and fill in the blanks.

Topic: Environment and you.

Google searched .... results in .... seconds.

Click on any one link and write five points about your search.

1.
2.
3.
4.
5.

So friends, in this unit we have learnt many new uses of computers like networking, Internet and internet safety, one of the most popular website/search engines, the Google. Now explore the world of internet yourself and take care about the safety issues.

ASSESSMENT EXERCISES

1. Tricky Terms:

   E I D.I A M L M T U

   E T R N I E N T

   P N O G O L Y

   W K R O E N T

2. Multimedia is a combination of .... with lot of graphics.

3. The Internet is a .... network of computers.

4. World Wide Web is a system of interlinked pages which display .... that can be accessed via ....

5. The internet can be used for getting ....

6. Give full forms of:

   ► ARPANET

   ► www

   ► LAN

   ► WAN

   ► E-mail

Unit End project:

🖰 Collect information regarding latest gadgets and gizmos from the internet.

🖰 Search the internet and find out five search engines similar to Google.

🖰 List five distinct features of Google.

🖰 List five anti-virus software available in the market.

CHAT: Simple safety steps for chat rooms: Let's dwell

Careful – people online may not be who they say they are.
Hang onto your personal information; never give out your e-mail or
home address, phone number or where you go to school.
Arranging to meet could be dangerous. Never meet someone offline
unless you are sure who they are, and then only in a public place,
with a friend/carer.
Tell your friends or an adult if you come across something that
makes you feel uncomfortable.

So friends, we have reached at the end of the topic "Computer". And now we are well aware about different aspects of it. But it is very important to take care of this amazing machine and its parts too. So in this final unit we will learn about the correct way to handle it.

▷ COMPUTER PARTS CARE

Cleaning the Machine

Like all other electronic equipments, computer also needs special care and attention in order to perform properly and safely.

Cleaning of computer may seem an easy task, but it requires special care during cleaning. The first step to computer **care is keeping it free from dust, dirt and** liquid.

Fig.1: A clean computer

Computer Trivia

Glass cleaner sprayed directly on the monitor could possibly leak inside the unit and cause damage. The glass cleaners are made up of strong chemicals and alcoholic spirit which cause the damage.

Monitor Care

Computer Monitor is a vital part of a computer and hence it needs some care.

The following tips will help us know how to secure your monitor from getting dirty and non-functional.

- Never spray any glass cleaner directly on monitor screen, instead spray a lint free cloth lightly with glass cleaner and then clean the screen.
- It should be kept covered when not in use so that no dust adheres on its screen and the back part.
- Monitors are extremely sensitive towards the influence of magnetic field so make sure not to keep any magnetic object near it, not even speakers and woofers
- Never try to lift the computer monitor by yourself.

Fig.2: Monitor

For glass CRT (television-style) monitors, use an ordinary household glass cleaning solution. Unless your manufacturer recommends differently, don't use alcohol or ammonia-based cleaners on your monitor, as these can damage anti-glare coatings. And never try to open the housing of a CRT monitor. Capacitors within can hold a dangerous electrical charge – even after the monitor has been unplugged.

For liquid-crystal display (LCD), laptop and flat-panel monitor screens, slightly moisten a soft, lint-free cloth with plain water. Microfiber cloths are excellent for this purpose. Avoid using paper towels, which can scratch monitor surfaces.

Do not spray liquid directly onto the screen – spray the cloth instead. Wipe the screen gently to remove dust and fingerprints. You can also buy monitor cleaning products at computer-supply stores.

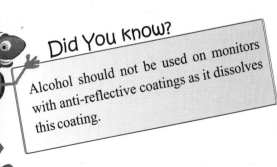

Did You Know?

Alcohol should not be used on monitors with anti-reflective coatings as it dissolves this coating.

CPU Care

It is important to remember that a computer is connected with electric wires. And the CPU is the main processing unit which is connected with a web of wires which carry high electric voltage.

It is advised that the wires should not be connected or disconnected when the electricity is on.

It is also advised that you should not touch the wires and attempt to plug them into electrical sockets.

Fig. 3: CPU

Keyboard Care

Keyboard is an essential and frequently used part of the computer, so we have to be very careful while using it.

As it has got many keys, the handling should be done more carefully.

- To clean your keyboard turn the PC off, then wipe off any dirt with a slightly damp cloth.
- Keep food and drink away from the keyboard. A spilled drink can quickly render a keyboard useless.

Fig. 4: Keyboard

Mouse Care

Just like keyboard mouse is also a delicate and frequently used part of the computer. So, it becomes sticky as it picks up dirt during use.

- Press the mouse buttons softly.
- Always keep mouse on a mousepad or a clean surface otherwise the scroll wheel or the optical surface may get dirty with dust.

ACCEPTABLE CLEANING AGENTS

- Plain water
- Vinegar mixed with water
- Isopropyl Alcohol (common "rubbing" alcohol) mixed with water
- Commercial products for this specific purpose
- A solution of Isopropyl alcohol and water can be used to disinfect and remove smudges, dirt and fingerprints from cell phones and PDAs.

Special Feature

When it comes to taking care of the computer, we do take care of the monitor, keyboard, CPU, mouse but we forget the printer. It is very necessary for the printer too to be taken care of which most users tend to avoid.

Position the printer on a sturdy and level surface. Allow sufficient space around the printer for air flow.

While cleaning your printer, always use a lightly dampened cloth to clean the outside of the printer. To clean the inside use a dry, lint-free cloth. This will ensure that no liquid enters the inner parts of the printer. Most manufacturers provide useful cleaning information on their manuals – always refer to them.

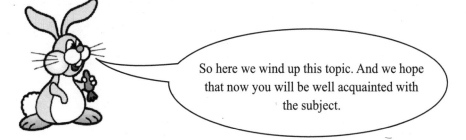

So here we wind up this topic. And we hope that now you will be well acquainted with the subject.

ASSESSMENT EXERCISES

1. Read the statements and correct them.

   a. We can lift the computer screen easily, from one place to another.

   ....

   b. In case of a system failure the CPU can be opened and checked easily.

   ....

   c. If sometime keyboard keys are not working properly we should press them hardly.

   ....

2. Fill in the blanks:

   a. Monitor screen should not be cleaned by _____.

   b. CPU wires carries _____.

3. Give reason in one line:

   a. Do not use alcohol or ammonia-based cleaners on your monitor.

   ....

   b. Press the mouse buttons softly.

   ....

Unit End Project

1. Arrange two keyboards, one properly functioning and one slightly non-functional. Now connect them to two separate systems and observe the difference in their operation.

Safety Measures: Let's dwell

The following cleaners should NOT be used on plastics:
Acetone: Familiar household uses of acetone are as the active ingredient in nail polish and spot removers and as paint thinner.
Ethyl alcohol: Best known as the intoxicating ingredient in alcoholic beverages. Also used for cooking and lighting, and as car fuel (Ethanol).
Ethyl acetate: Generally used as a solvent for cosmetics and food products. It is commonly used to clean circuit boards and is in some nail polish and stain removers.
Ammonia: A common household cleaning agent for glass, porcelain, and stainless steel.
Methyl chloride: A rubber solvent. Used as an additive in lubricating oils and motor fuels, and also used to clean up waterborne oil spills.

SELF-IMPROVEMENT/PERSONALITY DEVELOPMENT

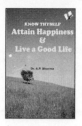

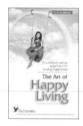

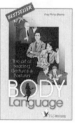

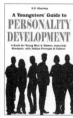

All books available at www.vspublishers.com

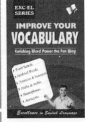

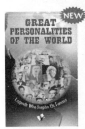

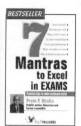

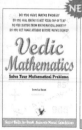

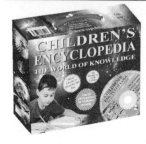
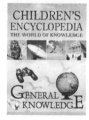
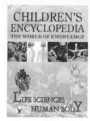
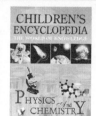
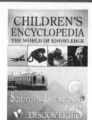

HINDI LITERATURE

MUSIC (संगीत)

MYSTERIES (रहस्य)

Also Available in Hindi

MAGIC & FACT (जादू एवं तथ्य)

TALES & STORIES

NEW **All Books Fully Coloured**

CHILDREN TALES (बच्चों की कहानियाँ)

All books available at www.vspublishers.com

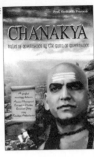

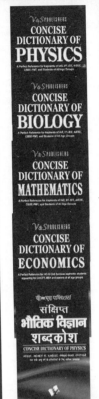

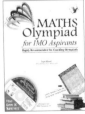